# ICME-13 Topical Surveys

**Series editor**

Gabriele Kaiser, Faculty of Education, University of Hamburg, Hamburg, Germany

More information about this series at http://www.springer.com/series/14352

Murad Jurdak · Renuka Vithal
Elizabeth de Freitas · Peter Gates
David Kollosche

# Social and Political Dimensions of Mathematics Education

## Current Thinking

Murad Jurdak
American University of Beirut
Beirut
Lebanon

Renuka Vithal
University of KwaZulu-Natal
Durban
South Africa

Elizabeth de Freitas
Education and Social Research Institute
Manchester Metropolitan University
Manchester
UK

Peter Gates
School of Education
University of Nottingham
Nottingham
UK

David Kollosche
Humanwissenschaftliche Fakultät
Universität Potsdam
Potsdam
Germany

ISSN 2366-5947          ISSN 2366-5955   (electronic)
ICME-13 Topical Surveys
ISBN 978-3-319-29654-8    ISBN 978-3-319-29655-5   (eBook)
DOI 10.1007/978-3-319-29655-5

Library of Congress Control Number: 2016931604

Printed on acid-free paper

This Springer imprint is published by Springer Nature
The registered company is Springer International Publishing AG Switzerland

Murad Jurdak · Renuka Vithal
Elizabeth de Freitas · Peter Gates
David Kollosche

# Social and Political Dimensions of Mathematics Education

## Current Thinking

Murad Jurdak
American University of Beirut
Beirut
Lebanon

Renuka Vithal
University of KwaZulu-Natal
Durban
South Africa

Elizabeth de Freitas
Education and Social Research Institute
Manchester Metropolitan University
Manchester
UK

Peter Gates
School of Education
University of Nottingham
Nottingham
UK

David Kollosche
Humanwissenschaftliche Fakultät
Universität Potsdam
Potsdam
Germany

ISSN 2366-5947          ISSN 2366-5955   (electronic)
ICME-13 Topical Surveys
ISBN 978-3-319-29654-8      ISBN 978-3-319-29655-5   (eBook)
DOI 10.1007/978-3-319-29655-5

Library of Congress Control Number: 2016931604

Printed on acid-free paper

This Springer imprint is published by Springer Nature
The registered company is Springer International Publishing AG Switzerland

# Main Topics You Can Find in This ICME-13 Topical Survey

This topical survey on the *Social and Political Dimensions of Mathematics Education* examines current thinking about issues in five critical social and political areas in mathematics education:

- Equitable access and participation in quality mathematics education: ideology, policies, and perspectives
- Distributions of power and cultural regimes of truth
- Mathematics identity, subjectivity and embodied dis/ability
- Activism and material conditions of inequality
- Economic factors behind mathematics achievement.

# Contents

# Chapter 1
# Introduction

This *Topical Survey on Social and Political Dimensions of Mathematics Education-Current Thinking* produced by the Topic Study Group (TSG) 34 is one of the series of the topical surveys associated with the TSGs of ICME 13.

The roots of the *Social and Political Dimensions of Mathematics Education* can be traced to the 1980s, to several seminal developments and publications (Vithal 2003), which gained so much momentum that a special fifth day was added to the ICME 6 programme in 1988, titled *Mathematics Education and Society*. Some 90 presentations were made by mathematics educators from diverse countries, which appeared in a UNESCO publication organized by Damerow, Bishop and Gerdes, and edited by Keitel (1989). This was immediately followed by the first conference on the *Political Dimensions of Mathematics Education* (PDME 1) with the theme of "Action and Critique" (Noss et al. 1990). The PDME 2 (Julie et al. 1993) and PDME 3 (Kjærgård et al. 1995) conferences were later replaced by the *Mathematics Education and Society* conferences, the first of which took place in 1998 (Gates and Cotton 1998) and have continued since then (see Further Readings). This first TSG 34 on the *Social and Political Dimensions of Mathematics Education* in ICME 13 is important in that it represents the mainstreaming of this area of work as a scholarly and ongoing significant activity of the broader mathematics education community.

Right from the start, the members of ICME 13 TSG 34, who are the authors of this publication, ruled out a conventional survey of literature on the social and political dimensions of mathematics education and opted to focus on what they considered five critical areas of the social and political dimensions of mathematics education, which are elaborated below. Furthermore, the team opted to focus mainly on current thinking in those five areas and only to go back in history as far as was needed to contextualize the current issues. As a result, the area of 'the role of economic and historical factors' was changed to 'economic factors behind mathematics achievement'. Each author took primary responsibility for writing one of the sections and for reviewing one section written by another author.

© The Author(s) 2016
M. Jurdak et al., *Social and Political Dimensions of Mathematics Education*,
ICME-13 Topical Surveys, DOI 10.1007/978-3-319-29655-5_1

This 'survey on the state-of-the art' of the social and political dimensions of mathematics education explores a range of issues within each of the five identified areas.

The first, titled 'equitable access and participation in quality mathematics education: ideology, policies, and perspectives', examines the issue of equitable access and participation in quality mathematics education in different contexts and from different ideological perspectives. It starts by identifying the ideological bases of equity and quality and how these are reflected in policies and practices as well as in the perspectives through which mathematics educators view this issue. The section also examines the attainment of the illusive, but sublime, goal of equitable access and participation in mathematics education in three political systems with different underlying ideologies: The USA as a liberal system, Cuba as a Marxist system, and Finland as a social democratic system.

The second, titled 'distributions of power and cultural regimes of truth', challenges the apolitical view of mathematics and mathematics education. It argues that through the systematic reproduction of socio-economically, ethnically and gender-based differences in achievement, mathematics education contributes to the development of inequalities in future opportunities for students. It goes further to ascertain the critical role of mathematics education research in addressing key concepts such as mathematical literacy or modelling. It concludes that the contributions on the political nature of mathematics itself provide new insights into the political bias of the mathematics in the classroom.

The third, titled 'mathematics identity, subjectivity and embodied dis/ability', explores current research on the political forces at work in identity, subjectivity and dis/ability within mathematics education, showing how emphasis on language and discourse informs this research, and how new directions are being pursued to address the diverse material conditions that shape learning experiences in mathematics education.

The fourth, titled 'activism and material conditions of inequality' traces the emergence and development of the notion of activism in mathematics education in the literature theoretically, in research, and in practice. It further points to connections between activism and material conditions of inequality. In particular, the notion of poverty is explicated as it has found expression in research across differently resourced contexts and especially large scale quantitative studies. While this has led to identifying "achievement gaps", other gaps such as "theory gaps" can be posited. Several issues and implications are explored including other domains such as curriculum reforms and the availability of advancing communication technologies.

The fifth, titled 'economic factors behind mathematics achievement', examines the political dimensions of mathematics education through the influence of national and global economic structures. By drawing on Programme for International Study Assessment (PISA) data it looks at patterns of underachievement and learning as connected to levels of social equity in a country and looks at how this might be understood. It further looks into the differential experiences of mathematics for pupils from lower socioeconomic communities and argues that this difference is not merely random or unimportant. Such differences of experience are systematic and structural with the result of further enhancing social inequity.

In the section on the 'summary and looking ahead', the results of our survey are presented. Based on the main findings, the topical survey looks ahead and suggests some ideas and research questions to help move forward the social and political dimensions of mathematics education.

# Chapter 2
# Survey on the State-of-the Art

## 2.1 Equitable Access and Participation in Quality Mathematics Education: Ideology, Policies, and Perspectives

### 2.1.1 Framing the Context of Equity and Quality of Mathematics Education

"Equitable Access to Quality Mathematics Education" reads and sounds like a political slogan, or at least, like a rhetorical statement. No one will contest its good intention; but almost everyone believes that this goal is an elusive, although very worthy goal. The connotation of the words 'equitable' and 'quality' overpower their denotation. However, the problem is exactly in what these notions denote. In the case of equity in education, debate centers on equity in *what* (access, input, processes, outcomes), for *whom* (students, schools), and *how* (policies, direct support, individual or community initiative). In the case of quality in education the issue concerns the meaning of quality and whether it should be applied to input, process, artifacts, outcomes, or other valued particular aspects such as social cohesion or universal aspects such as human rights.

At a most basic level, equity and quality issues in mathematics education arise when individual students engage in the collective activity of learning mathematics at the level of the classroom. To deal with issues of equitable access and distribution of quality in the mathematics classroom, the teacher has access to many possible practices—such as differentiation of instruction and developing high expectations of achievement from students. However, teachers' practices to promote equity and quality mathematics education are constrained, among other things, by school policies regarding reward structure, teacher professional development, improved technology, or attention to social circumstances. Thus teachers' practices in this regard are shaped, to a large extent, by school policies.

© The Author(s) 2016                                                                                    5
M. Jurdak et al., *Social and Political Dimensions of Mathematics Education*,
ICME-13 Topical Surveys, DOI 10.1007/978-3-319-29655-5_2

On the other hand, state policies shape school policies and hence have an influence over schools' ability to promote equitable access and distribution of quality mathematics education. State policies regarding school funding, school autonomy, school performance assessment are critical to the ability of schools to adopt practices that support equity and quality of the education they provide. The dilemmas are many in this regard; for example, if the state promotes between-school equity through funding schemes that may constrain school autonomy, it may risk compromising the quality of education by constraining schools' motivation to innovate (Jurdak 2014).

## 2.1.2  Ideology and State Policies in Relation to Equity and Quality

State policies have philosophical/ideological underpinnings but are not determined by them. This means that there may be a diversity of policies within the same ideological orientation. However, different ideologies tend to be associated with policies of different orientations. This section demonstrates how ideology mediates educational equity and quality policies by discussing three examples that represent the three ideologies of neoliberalism (USA), Marxism (Cuba), and social democracy (Finland).

The No Child Left Behind (NCLB) Act of 2001 in the USA focused on having all students proficient in reading, math, and science. All states had to develop learning standards and assessments of student performance. NCLB sets demanding accountability based-testing standards for schools, districts, and states with measurable adequate yearly progress objectives for all students and subgroups of students defined by socioeconomic background, race-ethnicity, English language proficiency, and disability. Individual schools were required to be on a path toward universal proficiency by 2014. Hursh (2007) views NCLB as a USA response to the growing competitiveness in the global economy, whose driving force is market capitalism. First, he argued the NCLB's emphasis on standardized testing was a means to provide 'objective' (as opposed to teachers' subjective assessment) measure of quality to the consumers (parents, schools, universities)—a neoliberal idea. Second, closing the achievement gap between advantaged and disadvantaged students by enabling parents to choose schools based on the objective measures provided by standardized tests, was not only meant to serve the social cause of equity but also to strengthen the ability of the country to compete in the global economy.

Cuba has followed a Marxist ideology since the 1959 revolution. Since then, and across the temporal shifts, political realignments, and a significantly changed world in terms of geopolitical power, Cuba was hailed for a consistent attainment of a high level of equitable distribution of quality education. A UNESCO study in 1998 on educational achievement in Latin America in 13 Latin American countries showed that Cuban students scored highest average which was 100 points above the regional average in mathematics (Gasperini 2000). Carnoy et al. (2007) show how the availability of work for school-age children and ready access to high quality public health care underpin student participation and performance in schooling. In addition, they cite the high quality teacher education, and a strong alignment between teacher education and school curricula, as general factors accounting for student performance. In terms of equity, Cuba had achieved not just universal primary schooling, but universal access, with unprecedented levels of equity, to pre-school, school, and tertiary education (Griffiths 2009). According to Griffiths, this achievement in equity is significantly linked to the policy of viewing schooling as preparation for work and linking the latter to national economic development.

Finland is an example of a country that is guided by social democratic ideology. "In the new millennium, Finland has gained a reputation for having one of the best education systems in the world" (Morgan 2014) as reflected in its high ranking in international comparison tests, its highly qualified and competitively selected workforce, and its non-competitive educational system. This is reflected in its superior welfare system, which offers, among other things, tuition-free education for all students and free early childhood care and health services. The Finnish strategy for achieving equality and excellence in education has been based on a 1972 reform in which a nine-year compulsory comprehensive system superseded the two-track system. The new system was a publicly funded comprehensive school system without selecting, tracking, or streaming students during their common basic 9-year education. According to Sarjala (2013), part of the rationale that led Finland to reform is determined by Finland's social values which include a devotion to equity and cooperation and which are reflected in the school system's ideology.

### 2.1.3 Perspectives on Equity and Quality in Mathematics Education

Educators and researchers in mathematics education have adopted a variety of perspectives to understand and study issues of equity and quality in mathematics education. These perspectives differ in their underlying philosophical/ideological underpinnings. As a result of surveying the recent literature, particularly the comprehensive book entitled *Mapping Equity and Quality in Mathematics Education* (Atweh et al. 2011) four distinct but not mutually exclusive perspectives are identified: the mathematical/pedagogical perspective, the socio-political perspective, the cultural historical activity theory perspective, and the humanistic ethical perspective.

### 2.1.3.1  Mathematical/Pedagogical (Pragmatic) Perspective

The mathematical/pedagogical perspective views quality and equity in mathematics education as issues that can be addressed within mathematics and its pedagogy. This perspective does not invoke any theory outside mathematics and its pedagogy to understand issues of equity and quality in mathematics education, and hence the name pragmatic. It recognizes the social context of mathematics learning as a 'given' which should be taken in consideration in designing and implementing pedagogical approaches to teach and learn mathematics. Implicit in it is the positivist assumption that there is a reality which is independent of human mind and that this reality can be modelled by mathematics, and that mathematics can be applied to understand this reality.

Quality of mathematics learning within this perspective is something that is defined by a community of users within the legitimacy of mathematics as a discipline. If there is a deficiency in the desired level or nature of quality of mathematics learning, then this deficiency can be addressed through appropriate pedagogical means. Similarly, any undesirable discrepancy in learning mathematics among individuals or groups can be redressed by additional pedagogical resources. The mathematical/pedagogical perspective is the dominant perspective in both research and practice. In practice, this perspective is dominant among teachers, schools, and governments. In research it encompasses all research that limits the framing and interpretation of research issues to mathematics and its pedagogy.

### 2.1.3.2  Socio-Political Perspective

A central concept in Skovsmose's theory of *Critical Mathematics Education* (Skovsmose 2011) is the relationship between mathematics, discourse, and power. Starting from the ideas of Michel Foucault, Skovsmose stipulates a relationship between power and language in the sense that power can be acted out through the applied language as a means of formatting reality. According to Skovsmose:

> If we combine the two ideas, i.e. that language is part of a formatting of reality and that language includes actions, then the way is opened for a performative interpretation of language and of the power-language interaction – and in particular with respect to mathematics. (p. 61)

Inspired by such a stance on critical mathematics education, many mathematics education researchers used this lens to study equity and quality issues in mathematics education. Pais and Valero (2011) argued that the inequity in mathematics education cannot be understood without understanding the relation between school and social mode of living—a classic Marxist position. Also, quality of mathematics education cannot be conceptualized without a critical understanding of the significance of valued forms of mathematical thinking within capitalism. In the context of technology-mediated mathematics education, Chronaki (2011) argued that self/society development through technology-related literacies is not merely a tool for better understanding mathematical concepts, but can be seen as a tool for introducing learners to certain standards of 'modern' life which is equivalent to the construction of a fixed 'rationality' as the ultimate goal for quality within the confines of imperialist, colonial and patriarchal discourses. Gutiérrez and Dixon-Román (2011) argued that the achievement gap-only discourse about inequity in the USA is not likely to liberate schooling from hegemonic practices but rather lead to viewing mathematics as a commodity that is sold to students while they are in school, which is very different from the way mathematics is used in society.

### 2.1.3.3   Cultural Historical Activity Theory Perspective

Leont'ev (1981) conceived of activity as a purposeful set of artifacts-mediated actions toward a desired object. Engeström (1987) formally introduced the collective activity as a system. The activity system is a collective activity consisting of a purposeful activity in which a subject (or subjects) is engaged to attain an object shared by a community of practice, using mediating artifacts, where responsibilities are assigned collectively among members of the community (division of labor) according to policies within the social cultural context (rules).

In activity theory, the idea of transformation is closely tied to the dialectic ontology of Marx and Engels. Transformation comes as a result of inner contradictions as humans engage in concrete activities in a dynamically changing world. Engeström (2001) developed the theory of expansive learning to explain learning of phenomena that, by their nature, cannot be identified ahead of time. Transformation occurs as a new learning both at the individual and collective levels as a result of appropriating these inner contradictions.

Jurdak (2009) has used the constructs of activity system and expansive learning to interpret equity and quality in mathematics education where he conceived of the lack of equity in a system as an inner contradiction in the division of labor of the activity system of mathematics learning and teaching, and the lack of quality as contradictions within the system that impede the attainment of the desired object of the system. It is through the dynamic process of conscious actions of individuals in the system that expansive learning occurs and thus dynamically makes the system balanced until new contradictions trigger a new cycle of transformation.

### 2.1.3.4   Ethical Social Justice Perspective

This perspective asserts that ethics and social justice are the core concepts for understanding quality and equity in mathematics education. According to Atweh (2011), who promoted the ethical social justice perspective in mathematics education, "Levinas constructs the encounter with the other as the bases of ethical behaviour. He posits the ethical self as prior to consciousness of the self, being and knowledge." (p. 72). From this perspective, the quality of mathematics education implies a responsibility (read response-ability) towards the other on the part of students to develop their capacity to transform aspects of their life both as current students and future citizens. Equity on the other hand is embedded in the broader concept of social justice which extends the responsibility for the other to the society that has many others. According to Atweh, social justice implies that dealing with individuals in isolation from their social group memberships, is unjust since it ignores the effect of a student's background on their participation in mathematics education.

In this section, the ideological bases of equity and quality and how that is reflected in policies and practices, are explored. The perspectives through which mathematics educators view equity and quality are also examined. The goal of achieving equitable access and participation in mathematics education, although illusive and sublime, seems to be achievable to a high degree in some countries such as Cuba and Finland.

## 2.2   Distributions of Power and Cultural Regimes of Truth

### 2.2.1   Linking Mathematics Education and Power

Mathematics education is a social institution which is inseparably linked to power. Mathematicians and scientists, education researchers, politicians, teachers, students and parents are interested in mathematics education for various reasons, for example for the recruitment of future specialists, for the education of the enlightened citizen, for the vitality of the state economy, for the pursuit of a meaningful and dignified purpose in life or for the allocation of beneficial opportunities in further education and work. The social influence of many of these cultural groups depends on the existence and legitimisation of mathematics education, while other cultural groups see their future social opportunities determined in the mathematics classroom. It is therefore obvious that mathematics education is shaping and itself shaped by various fields of socio-political interests. In the last few decades, these connections between mathematics education and the socio-political have become an object of critical research (Valero 2004).

The connections between power and education have been studied through different theoretical lenses. Especially sociological frameworks have been widely applied in mathematics education research on the topic, for example concepts such as ideology, alienation, groups of conflicting interests, reproduction of class differences and economisation as introduced by Marx (1972). While Marx understood social differences as determined by economic capital, Bourdieu (1986) also distinguishes cultural, social and symbolic capital, allowing a more differentiated view on the interplay between mathematics and power. Bernstein (1971) shows how different social groups use different codes of language and how the nearly exclusive use of the elaborated code of the middle class in school causes the reproduction of social inequalities, systematically hindering other students from educational success.

One of the most profound analyses of the interplay between power and knowledge was provided by Foucault (1984) who—drawing on sociology, philosophy and history alike—studied 'regimes of truth' which regulate what is accepted as true or rejected as false, how truth is acquired and who is legitimated to make these distinctions. Foucault's approach to always think knowledge and power together provides a language to critically approach commonly held convictions in mathematics education research and to understand mathematics itself as a regime of truth.

## 2.2.2 Reproduction of Differences

Mathematics education can be understood as a 'gate-keeper' deciding who is allowed or not allowed to pursue higher goals in education or profession. Stinson (2004) highlights how this selective function of education in mathematics can be traced back to Ancient Greece, where mathematics was considered an access to the essence of the cosmos. However, the basis upon which decisions on who is 'in' and who is 'out' are made is often ambiguous. Indeed, research shows that mathematics education systematically reproduced social differences based on socio-economic status, ethnicity and gender, often regardless of mathematical ability.

For example, students with low-economic status are systematically excluded from success in mathematics education by the wide-spread use of a language which is intelligible for high but misleading for low socio-economic status students. This has been thoroughly documented in studies which apply Bernstein's theory of language codes in pedagogic practice and focus on assessment (Cooper and Dunne 2000), school mathematics textbooks (Dowling 1998) and classroom interaction (Straehler-Pohl et al. 2014). Dowling also elaborates how the textbook for high socio-economic status students prepares them to become sovereign masters of mathematics whereas the textbook for low socio-economic status students does not support an understanding of mathematics but merely fosters the submissive recognition of the superiority of mathematical approaches. Discussing ability grouping in mathematics education, Zevenbergen (2005) describes similar mechanisms by the use of Bourdieu's notions of 'field' and 'habitus'.

As socio-economic differences are closely linked to ethnicity, the mechanisms described above have a profound impact on marginalized ethnic groups. Apart from that, these groups are faced with what Stinson (2013) calls the "white male math myth", a regime of truth ascribing mathematical intelligence to the white male population only. Martin (2009) discusses how ethnicity impacts mathematics education, for example through the abilities students of colour are assumed capable or incapable of by teachers, society and the students themselves, or through low financial and human support for the schools in their neighbourhoods. Similar studies have been conducted on the situations of Latino students in the USA (Gutiérrez 1999; Gutstein 2003) or Native Americans in south Brazil (Knijnik 1999). However, the research paradigms have shifted in the last decade. Gutiérrez (2008) criticised a "gap-gazing fetish" in mathematics education and argues that rather than comparing the performance of marginalized groups with that of white males, research should direct its attention to the ways in which students from underprivileged ethnic groups become successful in mathematics education, create their own mathematical identities and re-invent mathematics from their cultural background. Accordingly, younger contributions focus on success (e.g. Stinson 2013) and on re-writing social subjectivity (e.g. Valero et al. 2012). Similar shifts can be observed in gender research, where attention is redirected from a debatable deficiency in the achievements of girls compared to boys to the processes in which girls face constrains but also actively and often willingly construct their gender in the mathematics classroom (de Freitas 2008a; Walkerdine 1998; Walshaw 2001).

## 2.2.3 Preoccupations of Mathematics Education Research

Within mathematics education research, 'researching research' (Pais and Valero 2012) has become an attempt to reflect on the preoccupations of mathematics education research, allowing for alternative and self-critical approaches to the study of mathematics education (Brown 2010).

A considerable amount of contributions discuss the influence of educational policy on the design and assessment of mathematics education. For example, Lerman (2014) presents analytical tools based on Bernstein and Foucault to study the effects of political regimes of truth on mathematics teacher education. Brown et al. (2013) elaborate how TIMSS 'has changed real mathematics forever'. Kanes et al. (2014) describe mechanisms of mathematics educators positioning themselves within the regime of truth of PISA, while Tsatsaroni and Evans (2014) analyse PIAAC as a political contribution towards the governing of people through a total pedagogisation of society.

Other contributions directly address the assumptions underlying specific concepts in mathematics education: Popkewitz (2002) questions various discourses on the legitimisation of mathematics education. Llewellyn (2012) argues that in mathematics education, the concept of 'understanding' is used in either a romantic or a neo-liberal, functional interpretation, both obstructing teaching for social justice. Zevenbergen (1996) argues that a constructivist theory of learning gives a unilateral advantage to students with bourgeois background; and Radford (2012) builds on Marx and Foucault to challenge contemporary concepts of 'emancipation' in mathematics education. Pais (2013) uses ideology critique to question the assumption that mathematics has a use-value in mundane everyday activities, while

Lundin (2012) builds on psychoanalysis to suggest that mathematics education researchers suppress the apparent absurdity of most real life problems in order to sustain an ideology which renders their work and mathematics education itself meaningful.

### 2.2.4   Questioning Mathematics

Evans et al. (2014) analyse mathematical images in advertisements in British newspapers and show how on the one hand mathematics is used to enhance the trustworthiness of advertisements while on the other hand the extent of this use of mathematics severely depends on the targeted socio-economic group of a specific newspaper. As mathematics has ideological functions (for example emanating objectivity), it cannot be considered apolitical and deserves a thorough analysis of the distributions of power and regimes of truths connected to it. While such a critique of mathematics has already been approached from a philosophical and historical perspective (Davis and Hersh 1983; Porter 1996; Desrosières 1993), research in mathematics education has only in the last decade begun to question the myth of a neutral 'pure' mathematics.

In a recent German study, Ullmann (2008) explores how the philosophically problematic and often criticised "myth" that Western mathematics was "secured, true, rational, objective and universally valid" (p. 11, our translation) is constantly reproduced throughout society, especially in the mathematics classroom. He argues that the myth of mathematics serves as an ideology, allowing mathematics to play the role of a trustworthy mediator between political or intellectual ambiguity and the Modern quest for objectivity. de Freitas (2004) provides valuable insights in the tensions and problematic which such a view of mathematics produces in teaching.

Mathematics is increasingly perceived as a negotiable field of social practices which arose out of specific needs, serves certain interests and implies various possibilities and restrictions for the perception, understanding and shaping of our world. For example, Radford (2003) understands algebraic symbolism from a cultural-historical perspective, arguing that it represents a new form of language that, emerging in the Renaissance, allows new representations of knowledge, thus linking Modern mathematics to Modern thought in general. Kollosche (2014) draws stronger connections between socio-political dimensions of mathematics and contemporary education by analysing Aristotelian logic and Modern calculation as a form of intellectual conduct and a regime of truth, which is 'democratised' in the mathematics classroom and allows for the organisation of Modern society. Ongoing research is analysing how far Euclidean geometry distorts naive perceptions in order to create a cohesive mathematical model of space (Andrade and Valero in press). Eventually, de Freitas (2013) proposes a new conception of mathematics, focusing on mathematics events rather than on mathematical objects in order to gain a theoretical grip on the social situatedness of and learning processes in mathematics.

## 2.3 Mathematics Identity, Subjectivity and Embodied Dis/ability

### 2.3.1 The Lived Experience of Mathematics Education

During the last three decades, research in mathematics education has turned to the concepts of identity and subjectivity as a way of studying the *lived experience* of mathematics education. This research sheds light on how students, teachers, parents, as well as mathematicians and other professionals, invest in particular kinds of identity or subjectivity in relation to mathematics. While some researchers have shown how students who think of themselves as creative don't identify with the discipline of school mathematics (Boaler and Greeno 2000), others have shown how masculinity intersects with mastery identities associated with mathematics achievement (Mendick 2006; Solomon 2009). Socio-cultural and socio-political approaches to identity have emphasized the fragmentation, multiplicity, contingency and partiality of identity, allowing for non-essentialist studies of how people do and do not identify with mathematics (de Freitas 2004, 2008; Walkerdine 2004). Research methods in this area look to how particular discursive practices offer evidence of how students are positioned (and how they position themselves) within school mathematics.

Research on teacher identity in mathematics education has emphasized the complex conflicted affiliations of teachers working within a high-stakes discipline (Drake et al. 2001; Walshaw 2004, 2013). Much of this work focuses on the ways that teachers develop policy-inflected identities in relation to various managerial, professional, and global reform discourses (Rodríguez and Kitchen 2005). Research methods in this area tend to rely on interviews, questionnaires, observation, and ethnography, focusing on how teachers' language-use and behavior in classrooms reflect various identifications.

### 2.3.2 What Is Identity?

Benwell and Stokoe (2006) use the term identity to simply designate "who people are to each other" (p. 71). They point to the instances when a speaker explicitly invokes a relationship category, such as "I am her sister", accomplishing locally relevant conversational goals. But identity is enacted in various ways, not always explicitly. Tracy (2002) differentiates between "master identities"—those defined in terms of race, class, gender, sexual orientation, and other tags in current circulation—and "interactional identities" that emerge during moments of interaction, what one might refer to as "discourse identities" or "situational identities". Teacher acts of positioning and marking of identity (such as "You have to listen, because I'm the teacher" or more subtle markers of identity work such as "Ok, let's hear from someone who is putting this all together?") may seem to *interrupt* what is taken to be the goal of instruction, or at least offer a slight diversion, but they can also function as discursive moves that *further the agenda* of education goals (good and bad) while strongly positioning those involved in the interaction. Wetherell (1998), for instance, locates explicit identity statements within the broader institutional context and in relation to broader regimes of governance. On the other hand, identity work is complex, and attention to students' actions at the micro-scale, perhaps documenting the way that gestures or facial expressions communicate tacit identity markers, "provides in rich technical detail how identities are mobilized in actual instances of interaction" (Widdicombe 1998, p. 202).

The concept of identity has often been considered problematic because it is often used in research to capture *essential* characteristics of people. Without intending to do so, some research on identity in mathematics education seems to re-entrench stereotypes about what sorts of identities excel at mathematics. In an article entitled "Who needs identity?" the cultural theorist Stuart Hall (1996) suggests that the term identity remains a useful theoretical construct precisely because it is so thoroughly "under erasure". In other words, it is precisely because identities are fluid and open to change—and identifications are typically *partial*—that the term has value. For Hall, identity remains a crucial theoretical concept because it functions as a site for questioning the nature of affiliation, and forces us to confront the way that power relations are lived.

> Precisely because identities are constructed within, not outside, discourse, we need to understand them as produced within institutional and historical sites, within specific discursive formations and practices, by specific enunciative strategies. Moreover, they emerge within the specific modalities of power, and thus are more the product of the marking of difference and exclusion, than they are the sign of an identical, naturally constituted identity – an identity in its traditional meaning (that is, an all-inclusive sameness, seamless, without internal differentiation) (Hall 1996, p. 4).

### 2.3.3   Structure, Agency, and Subjectivity

Another challenge for research on the political dimensions of identity is assuming a binary between structure and agency. In other words, this binary creates a simplistic image of power distribution: identities are considered the site of potential agency, either seen as compliant or resistant to structures that are considered hegemonic and static. This work tends to locate the power to change at the level of the individual human subject, and denies this power to larger structured collectives— say institutions. Wortham (2006) argues that the study of identity needs to be wary of this binary between structure and agency—we should be wary of studying identity as the site of agency in relation to some inflexible behemoth of mathematics education. He warns that such an approach sets up too strong a binary, and excessively constrains what one is able to see as a researcher. An overly simplistic image of mathematics or school mathematics as 'big brother' creates a simplistic image of resistance and agency.

Recent shifts in the study of mathematics student subjectivity have emphasized the power of the students beyond a simple resistance model. Gutiérrez (2013) and Esmonde (2012), for instance, are critical of research that emphasizes the "achievement gap" of Latina and African-American students because of the way it actually entrenches deficit identities; they argue instead that we must explore how these students' different identities can be affirmed from *within* mathematics education.

Research on *subjectivity* in mathematics education tends to focus on the ways that mathematics education is lived at the level of the personal individual human, again focusing on discourse and language-use (Valero and Zevenbergen 2004). Theorists like Michel Foucault and Mikhail Bakhtin are often evoked in research on mathematics subjectivity, because these theorists offer important tools for analyzing power and discourse (Black et al. 2011). Research on subjectivity often draws on ideas from psychoanalysis, especially the work of Jacques Lacan. Pais (2013), for instance, draws on Lacan to show how ideology works through the global policy-speak of mathematics education, arguing that student and teacher subjectivity are entirely constrained by the force of ideology. Brown and McNamara (2011) use Lacan to show how identity and subjectivity entail never-ending attempts to tell stories about the self that adequately capture it, but that such stories are never adequate nor satisfying to the teller. The Lacanian approach claims that this misrecognition is in fact the engine of identity: "the stuff of personal construction is an attempt to reconcile one's view of oneself with the views one supposes others have of you… For Lacan, it is the gap that defines identity" (p. 100). In their research on pre-service elementary teachers, they show how the stories went from mathematics is "scary" to eventually a far more comforting identification with the rhetoric of the national policy about numeracy. Pre-service teachers began to speak the numeracy policy without question, stating "It's sort of ingrained into my head" (p. 76).

## 2.3.4 Dis/ability and the Body

We have included embodied dis/ability in this section of the essentials document because the production of bodies as mathematically abled or disabled intersects with other kinds of identity (race/class/gender/sexual orientation/geography/etc.). Moreover, the specific ways that bodies are impacted by mathematics education is all too often overlooked in identity/subjectivity research, despite these obvious intersections. Recent research on the political dimension of mathematics dis/ability problematizes the image of the ideal mathematics body with particular kinds of perceptual capacities and neurocognitive tendencies (de Freitas and Sinclair 2014). There is a growing awareness of the way in which certain sensory modalities are privileged, often only implicitly, in school mathematics. There are, of course, obvious ways in which assumptions about abled bodies (i.e. sight and hearing) are embedded in classroom practice, such as the privileged role of the blackboard and the extensive use of verbal instructions and explanations in Western classrooms. The current emphasis on alpha-numeric competencies in US mathematics education reflects another bias with implications for who and how dis/ability is produced. These curricular emphases reflect particular political historical investments into particular kinds of mathematics and impact differently abled bodies everywhere within the global market of education policy.

School mathematics is formatted by certain sensory assumptions about mathematical knowledge and these are tacitly incorporated into what it means to be a good mathematics student. The *practices* of school mathematics entail *material embodied habits* of entrainment. Thus identity and subjectivity are produced at micro-scales, through small often unexamined bodily practices that are taken for granted as part of 'classroom participation'. These micro-habits of participation are bound up with particular kinds of mathematics curricula. When curriculum focuses more on cardinality rather than ordinality, for instance, particular kinds of ability are validated. In her book *Teaching Mathematics to Deaf Children*, Nunes (2004) shows how deaf children often underperform on counting tasks, but their more spatial way of thinking of number makes them better than hearing learners on ordinal tasks such as counting backwards and "What comes after $x$?". These other tasks, however, are not the tasks that are used to establish socio-mathematical norms around number sense. Healy et al. (2011), who have explored how tactile means of learning about 3D shapes enable blind students to learn mathematics, showed how what is taken to be a mathematical abstraction might be very different when it is expressed through moving hands. In their chapter in the *International Handbook of Mathematics Education*, Healy and Powell (2013) draw on Vygotsky's stance on organs (the eye, the ear, the skin) as tools that can be considered in much the same way as material instruments, language or other semiotic resources. Increased focus on the role of the body in teaching and learning is opening up the discussion as to how particular kinds of bodies are hailed and indeed produced through particular mathematical practices (de Freitas and Sinclair 2014). We raise these concerns here because it seems essential that we pay attention to aspects of identity and subjectivity that are 'embodied' in different ways, and not always enacted through discourse and language. As researchers turn increasingly to neurocognitive science to identify sources for mathematics dis/ability (i.e. "dyscalculia"), it becomes increasingly essential that we critically examine what kinds of identities and subjectivities are produced through this research (Gifford and Rockliffe 2008).

Borgioli (2008) argues that learning dis/abilities in mathematics are constructed using a narrow definition of both what counts as acceptable mathematics and what counts as evidence of mathematics proficiency. In practice, many students labelled with a mathematics learning disability receive highly directed step-by-step instruction in rote learning of basic skills and procedures (Baxter et al. 2002; Fuchs et al. 2002). This focus persists whether these students are pulled out for special education interventions or offered differentiated instruction within regular classrooms (Hehir 2005; Woodward and Montague 2002). Typical kinds of instructional strategies offered for learning disabled students—such as grouping similar problems together for the sake of easy recognition, or coding operations in colour for easy association—reveal the assumptions about mathematics that are at work in these approaches. As researchers turn to the study of the body and other material dimensions of school mathematics, political questions about the nature of identity and subjectivity are newly posed.

## 2.4  Activism and Material Conditions of Inequality

### 2.4.1  Activism in Mathematics Education

The theme of activism and material conditions of inequality, as part of the social and political dimensions of mathematics education, may be considered to be a continuation of the conversation begun by the ICME 12 survey team on "Socioeconomic Influence on Mathematical Achievement: What is Visible and What is Neglected" (Valero et al. 2015).

For an overview of the literature it is useful to return to foundational writings in mathematics education where a notion of activism was initially formulated substantially and theoretically. In *The Politics of Mathematics Education*, a first volume that explicitly set out what it means to make a political reading of mathematics education, Mellin-Olsen (1987) paid particular attention to students who fail to learn mathematics, and interpreted that failure as political because some are denied access to the "thinking tools of the curriculum". He argued that mathematics is consciously resisted by students who reject the subject; and that students lack the appropriate meta-knowledge to engage the conflicting messages of school mathematics. Following this line of thinking, students enact an implicit activism in choosing to engage (or not) mathematics teaching and learning.

The notion of activism in mathematics education has its theoretical base in critical perspectives in mathematics education. Research and theoretical expositions in critical mathematics education have been particularly important in this respect through the seminal writings of Skovsmose (1994) and have continued to be developed.

> At the core of our work in exposing mathematics education as an inherently political enterprise is the dialectic between reflection and action... We use the term "critical agency" to express the dialectic between action and reflection. (Greer and Skovsmose 2012, p. 6).

In this way notions of activism in mathematics education are linked to awareness, reflection, action and agency.

These ideas have found expression in practice in a number of forms—inquiry or problem oriented; project-based; realistic or real-world mathematics education approaches and so on—that variously attempt to connect mathematics education to society to generate awareness and action about the myriad of ways in which mathematics education functions in society to include or exclude. It has led to a variety of research approaches; and it has also led to a focus on particular groups of students and teachers, arguably mainly at the margins—those who fail to learn, or are at the periphery, such as particular gender, ethnic, minority or socio-economic class.

This has also led to a much more politically explicit mathematics education. Gutstein (2003, 2012), for example, has demonstrated how mathematics teaching and research in schools described as "economically battered in a Black (African American) and Brown (Latina/o)" can engage students

> in the complexities of reading their world with mathematics and, to write the world with mathematics, they shared what they learned with their community and others in public presentations and through actions. (Gutstein 2012, p. 24).

Mathematics education is directly intended and enacted to produce an activism. The complexity of these approaches resides in what is deemed the margin of some contexts, and are the centre of others in that they constitute the majority. Hence what this activism means and how it gets expressed may well have very different implications and consequences.

### 2.4.2   Material Conditions of Inequality in Mathematics Education

It is understandable that practices and studies that advance activism in mathematics education dominate in contexts of schooling that are lacking in some respects (particularly resources), precisely because they seek to empower participants, draw attention to inequalities and inequities in educational opportunities and outcomes, and to act on addressing these.

In recent years, arguably, it has been the large scale quantitative studies that have brought greater visibility to actual inequalities in the material conditions in which mathematics teaching and learning arc taking place and results in differential outcomes. In their study of twenty years of TIMSS data, Vijay et al. (2015) "show that there is a strong correlation between the Human Development Index (HDI—a composite measure that captures both economic and social development aspects such as life expectancy, average years of schooling and GDP per capita) and mathematics achievement based on 2011 TIMSS data, a higher HDI is related to higher levels of achievement in mathematics." (p. 2). A persistent "achievement gap" has been identified in test driven quantitative studies for particular groups of students variously defined in different contexts along dimensions such as socio-economic class, minority, gender, ethnic, race or language groups.

The inequalities in the material conditions within which mathematics education takes places can be reflected on a continuum in the kinds of schooling from affluent to resource poor. The notion of 'poverty' is gradually emerging in the literature to better understand, and to develop interventions, policy and practices for teaching and learning mathematics in poor schools and for learners from disadvantaged backgrounds (e.g. Graven 2014; Lee 2012; McKinney and Frazier 2008; Balfanz et al. 2006; Kitchen 2003; Turner 2000; Payne and Biddle 1999). A number of these studies focus on what are referred to as "high-poverty schools", especially in the USA, and the implications of the introduction of various mathematics standards and educational reforms.

The very notion of poverty, however, takes on different meanings, and refers to widely differing material conditions. In affluent Western countries a mathematics classroom in a 'high poverty' school may be described as dilapidated. When compared to, say, poorer African countries, poverty refers to much more extreme conditions where the very structures of a classroom (roof, walls, doors, desks) or basic conditions (water, electricity, sanitation) may not exist. In this context it is not only the activism of students that is important but also the activism of practitioners, researchers and policy makers to draw attention to and influence policy and resource allocation within mathematics education. In this way activism and material conditions of inequality in mathematics education are connected in this theme in engaging the social and political dimensions of mathematics education.

### 2.4.3 Some Current Issues and Questions

Recently considerable activism has been observed in many different parts of the world for different reasons including political ones as well as those related to inequalities in material conditions. Critical and socio-political perspectives in mathematics education research, theory and practice have increasingly advanced the case for activism. The question however, does arise: to what extent or in what ways (if any) can the activism observed in any instance be linked to the mathematics education curricula or practices in any sense or form? Are the theoretical propositions and claims for activism through mathematics education sufficiently developed or are they exaggerated?

This is not to discount numerous studies reporting on social and political awareness and even action by learners through mathematics education in local settings of schools and immediate communities. The questions are asked to interrogate the veracity of the theoretical and conceptual propositions in expounding of social and political dimensions of mathematical education at a societal level. What can or does such a mathematics education mean if enacted in contexts where the majority receive that education in conditions of extreme poverty and considerable inequality?

A critical or socio-political mathematics education is understandably argued for particular contexts—schools and students described as being variously disadvantaged or poor. A case less commonly made is for equally developing such awareness and concern also in more affluent schools and classrooms to create caring and ethical societies; and also because it is often students from these very contexts that populate positions of leadership and influence in societies. If such a thesis is accepted then how is a critical, socio-political or social justice oriented mathematics education to be advanced across settings? Can such a mathematics education put students in harms' way when they challenge the authorities in real ways; and if so what ethical and moral dilemmas does this raise for mathematics education practitioners, researchers or policy makers?

While it is argued that the focus on material conditions of schools and mathematics education has, in part, been brought to fore by large scale quantitative studies, it has also created a particular discourse that has been, ironically, on the one hand, damaging to mathematics education in how it reduces mathematics education to competitive scores and league tables; and yet on the other hand draws public attention to and produces action on addressing inequities at a system level. Even though there may be lack of agreement on the interventions, these studies shape policy and dominate the political terrain in which decisions are made that directly impact mathematics education on the ground. Such studies have drawn researchers from multiple disciplines such as economics, statistics, development and policy studies; and it could be argued have in some cases complemented or in others displaced mathematics education research(ers) (Greer 2012). It raises questions about what research and whose research influences and shapes mathematics education—systemically in practice and at a system level?

While the notion of an "achievement gap" has gained currency and led to greater attempts to research and address inequities, it has equally also been critiqued. Lubienski and Gutiérrez (2008) critique "gap gazing" and highlight the benefits and dangers of focusing on the educational differences between groups and supporting any one group in discussion on equity in mathematics education. Turning the "gap gaze" back onto mathematics education, does however, open for other questions about research and theoretical gaps in studies on activism and material conditions of inequality; specifically, about the lack of theorisation and investigations on, for example, activism in and through mathematics education: in large scale meta-analysis or macro studies; in extreme conditions of poverty, in deep rural or conflict ridden contexts which dominate in poorer countries; or beyond schooling in tertiary and higher education generally.

What questions are researched, by whom, and for what purposes in mathematics education shapes and frames theoretical discourses as they emerge, get taken-up and are developed (or disappear). A "theory gap", arguably, seems to also have persisted in mathematics education research with reference to the themes of activism and material conditions of teaching and learning mathematics. Pais and Valero (2012) provide an insight into this issue by making a "distinction between what has been called a socio-political turn in mathematics education research and what we (they) call a positioning of mathematics education (research) practices in the Political" (p. 9). They examine the role of contemporary theories in mathematics education and in how "objects" of research (such as 'learning' or 'mathematics') come to be constructed. This may be extended to choices in research "subjects" and contexts and might well explain why mathematics education research and theorisation has not developed in, from or applied to the material conditions in which it takes place when this is the case and relevant for the vast majority of learners globally, and in the poorer and most populous countries. How are these silences to be addressed since these conditions have persisted over generations as teachers and learners continue with their mathematics education whatever and despite the conditions?

## 2.4.4  Implications for Other Domains

The themes of activism and the material conditions of inequality are particularly relevant in contexts where curriculum reforms are taking place. Numerous countries are engaging in changes to their mathematics curricula in a globally connected world.[1] Varying factors account for an increasing convergence of mathematics curricula across countries. However, such official or national curricula often belie the significant disparities in the conditions for teaching and learning mathematics. The weaker and most fragile parts of the education system, which coincide with those that have the poorest material conditions, are often also most negatively impacted by changes in curricula (Vithal 2012).

It is well established that one of the most important factors that shape differential mathematical learning outcomes is the quality of teaching and teacher education. Differentials in teacher knowledge and skills and their distribution in a mathematics education system parallel the material conditions in which mathematics education is delivered. A significant knowledge base has risen in teacher education about the optimal kind of knowledge needed to deliver a quality mathematics education, but more controversially, much less is known about the minimum needed to ensure that a basic mathematics education is delivered by teachers, especially in systems in which large proportions of teachers are un- or underqualified.

The rapid rise and pervasive availability and use of information and communication technology has brought a completely new dynamic and complexity to mathematics education practice and research. It is a development that has potentially undermined the material and resource inequalities and divides of society, as for example, seen in the massive spread of mobile phone technology; and yet it has also simultaneously widened the gap in who has access to which recent and more powerful technologies and connectivity. In mathematics education it has changed the ways in which the youth in particular, access information and enact an activist role in far more impactful and immediate ways then could have been imagined in the past. What this can or does mean for mathematics education practice, research and theory for the future is very much an open question.

---

[1]See http://www.mathunion.org/icmi/activities/database-project/introduction.

## 2.5 Economic Factors Behind Mathematics Achievement

### 2.5.1 Mathematics for Some

Socio-economically advantaged students and schools tend to outscore their disadvantaged peers by larger margins than between any other two groups of students. (OECD 2013a, p. 34)

This is indeed a really sobering thought, that economics, income inequality or social economic status (SES) is more significant in explaining differences in mathematics achievement than gender and race. Whilst this might cause some unease, it is just obvious. Being a poor student does not mean you can't go on to do well, earn a salary significantly over the median wage, or even write an article for an international conference! But while some students from disadvantaged backgrounds can succeed "against the odds" (Bembechat 1998), the system leaves many where they are. Social mobility becomes the story of the few not the many; history is written by the winners.

In "*Is* mathematics for all?" Gates and Vistro-Yu (2003), argued that across the world indeed mathematics wasn't for all, but was differentially experienced. They suggested several strategies to help a process of democratisation of mathematics: detracking, equitable allocation of resources, and the appreciation of working class cultures. A decade later and we are still arguing for the same strategies, which begs the question—why? Something must be going on to sustain the levels of inequality within the teaching and learning of mathematics in the face of much apparent consternation and displeasure. What are we doing wrong, or rather not doing right? Or maybe, more sinisterly, is this inequality sustained because it is what some desire.

This section starts with a focus on the Organisation for Economic Co-operation and Development (OECD) analysis of the Programme for International Study Assessment (PISA) 2012 data and in particular what that says about poverty and achievement: across OECD countries, a more socio-economically advantaged student scores the equivalent of nearly one year of schooling higher than a less-advantaged student (OECD 2013a, p. 13). What the PISA studies consistently show is, at the national, school and individual level, SES is clearly associated with mathematics achievement in a complex way (OECD 2013a, p. 37). However, what current economic and social analyses are showing is that it is not the existence of *poverty* itself that is the result of many social problems, but the existence of income inequality.

> The highest-performing school systems are those that allocate educational resources more equitably among advantaged and disadvantaged schools and that grant more autonomy over curricula and assessments to individual schools. (OECD 2014, p. 4).

So whilst the mathematics education research community might want to frame the debate on mathematics achievement around cognitive development, identity, curriculum, teaching style etc. we are up against a much bigger problem—growing income inequality. "A growing body of evidence points to high and rising inequality as one of our current decade's most important global issues" (Stotesbury and Dorling 2015, p. 1). The extent of the malignant effect of inequality has been well illustrated (Wilkinson 2005; Wilkinson and Pickett 2010)—greater equality increased everyone's quality of life. But if we want social class to have less influence on educational (and therefore mathematics) outcomes, "it will be necessary to reduce the material differences which are so often constitutive of the cultural markers of social differentiation" (Pickett and Wilkinson 2015, pp. 323–324). However, whilst there is a very strong tradition of mathematics education research situated within a social justice framework, as mathematics educators we surely need to be prepared to argue for the inevitable conclusion—to reduce the wealth of the affluent and distribute it to the poor. This will be a difficult process for many in the mathematics education community, yet is exactly what has been proposed by Wilkinson and Pickett (2010, p. 108) for some time.

Drawing together data from a range of international sources, Stotesbury and Dorling (2015) examined the PISA data on mathematics achievement. Their analysis suggests two things. First mathematical achievement is negatively correlated with income inequality (as measured by the ratio between the wealth owned by the top 10 % and the bottom 10 %), but second this correlation is significantly stronger when we measure the mathematics ability of older (16–24) students.

> This is interesting because it hints at the possibility that more unequal countries' education systems fail to foster long-term understanding to the same extent that education in more equal countries appears to have a longer lasting effect on young peoples' ability. (Stotesbury and Dorling 2015).

In other words, in countries with low levels of income inequality, what is taught in school mathematics seems to be retained longer once the student leaves the education system than where income inequality is higher.

That is a quite a surprising claim. How might a macroeconomic statistic on measures of relative wealth influence how students learn mathematics even after they leave school? Well the PISA data would need to be mined in a lot more detail to uncover the causal mechanisms at work. Dorling's work and that of Wilkinson and Pickett point to a number of social characteristics of unequal societies—increased social conflict, anxiety and insecurity, homicide, etc. and it is to these where we may find some of the root *causes* (but not the *manifestations*) of underachievement in mathematics.

In a study of mathematics education in "high performing" countries, Askew et al. (2010) argue that attainment in mathematics might be "much more closely linked to cultural values". This they admit "may be a bitter pill for those of us in mathematics education who like to think that how the subject is taught is the key to high attainment" (p. 12). Yet the way we respond to that may also be both cultural, and, importantly, political. Askew et al. argue that no system has the definitive answer; that choices need to be made between some very central social characteristics. So "you can have an egalitarian education and high standards (Finland), or you can have a selective one and still have high standards (Singapore)" (p. 14). The question is though, whose choice is it and how is that choice made? The economic and political system itself facilitates some choices over others. Yet as researchers we too have choices. The word "politics" does not appear in Askew et al. who prefer a focus on "socio-cultural-historical backgrounds". What Dorling and Wilkinson and Pickett's work offers us, is a *different* choice of emphasis that complements the focus on characteristics more visible in mathematics education. The single reference in Askew et al. (2010) to "(social) class" (remember *the* most significant characteristic according to the OECD) is restricted to a discussion of how parental social class in China is not a significant discriminator when looking at parental expectations. (See Gates and Guo (2013) for a discussion of the influence of social class in British-Chinese student achievement).

But it does not have to be like this, "countries do not have to sacrifice high performance to achieve equity in education opportunities" (OECD 2013a, p. 3), "Mexico, Turkey and Germany improved both their mathematics performance and their levels of equity" (OECD 2013a, p. 26). The OECD analysis also illustrates that merely increasing expenditure on education will not bring about improvement in achievement if it is not accompanied by greater equity. It is not a matter of how much is spent, but on how it is spent.

> In particular, greater equity in the distribution of educational resources is associated with higher mathematics performance. 30 % of the variation in mathematics performance across OECD countries can be explained by differences in how educational resources are allocated between advantaged and disadvantaged schools. (OECD 2013a, p. 29).

In highly differentiated educational systems, the impact of a student's socio-economic status on his or her educational goals is stronger than in less differentiated systems (Buchmann and Dalton 2002; Buchmann and Park 2009; Monseur and Lafontaine 2012) because

> socio-economically disadvantaged students tend to be grouped into less academically oriented tracks or schools, and this has an impact on their educational aspirations, possibly because of the stigma associated with expectations of lower performance among students enrolled in these tracks and schools, or because less – and often poorer quality – resources are allocated to these schools. (OECD 2013b, p. 86).

### 2.5.1.1 Mathematics Isn't for All

We have had "*the social turn*" (Lerman 2000) and the "*socio-political turn*" (Gutiérrez 2013) both of which encompassed a view of inequity as "a problem affecting particular groups of people [rather than] a problem of the school system" (Pais 2014, p. 1086). As a result of an individualisation of failure, attention has been directed away from the economic system, which by design creates inequality. It might also explain why the systematic failure of children from working class communities gets so easily overlooked, despite all the research that has explored this area (Pais and Valero 2012); and, taken further; it is because of the "*depoliticisation* of research" (Pais 2014, p. 1090) that allows research to cast a blind eye to the most significant source of underachievement. One source of this depoliticisation is a tendency to assume, that postmodern approaches offer insights because they "move beyond Marxist views of power" (Gutiérrez 2013, p. 12). Given the power of Marxist analyses of the economy, such individuating constructs as discourse, identity and a focus on localised struggles, whilst locally useful, can only fail to grapple with the structural inequalities which are an inherent component of international capitalism.

To understand the differential performance of pupils from low SES backgrounds, we need to look into classroom practices to ask difficult questions about the experiences of learners from certain social-economic groups. Much literature in the field of mathematics education focuses on teaching and learning and on levels of pupils' attainment through a focus on the pupil, the classroom, the teacher, the curriculum, and the school—in other words on the localised manifestation of cultural practices. Fewer studies drill down into the very structure of the economic and political system exploring how it solidifies into the interactions and artefacts of mathematics education.

There are some robust examples of inquiries into social class, one such is Lubienski's study of the mathematical experiences of pupils (Lubienski 2000a, b, 2002). Whilst she expected to find SES differences, what she found were very *specific* differences in whole class discussion and open-ended problem solving. High SES pupils thought discussion activities were for them to analyze different ideas whilst low SES pupils thought it was about getting right answers. The two groups had different levels of confidence in their own type of contributions with the low SES pupils wanting more teacher direction. Higher SES pupils felt they could sort things out for themselves. Here, social class is a key determining characteristic which can shed light on studies of discussion based mathematics.

A second area where Lubienski (2000b, 2002) noted differences was open-ended problem solving. The high level of ambiguity in such problems caused frustration in low SES pupils causing them to give up. High SES pupils engaged more deeply. It is well known that middle class pupils come to school armed with a set of dispositions and forms of language which are exactly the behaviours that teachers are expecting and prioritise (Zevenbergen 2000). High SES pupils have a level of self-confidence very common in middle class discourses, whilst working class discourses tend to be located in more subservient dependency modes (Jorgensen et al. 2014). So how does all this happen? Middle class pupils tend to live in families where there is more independence, more autonomy and creativity (Kohn 1983). The middle classes grow up to expect and feel superior with more control over their lives. Class is never far away from the mathematics classroom, but it is often far away from mathematics education research.

## 2.5.2   *The Poverty of Experience*

What is different for diverse pupils is the form that school mathematics takes. Some pupils will remain within a somewhat abstract world where the systems of thought of the school will be exactly what they need to move onto a next stage—be it further study of mathematics or higher education. For others however, those whose trajectory will be moving toward employment in some form, their school mathematics will be at odds with what everyone knows is needed to practice. Recent work on workplace mathematics has shifted a focus away from a conceptual, cognitive approach to a more situated and cultural approach (Hoyles et al. 2010; Roth 2014). Not only has this changed the way we see mathematics in use, but it has also contributed to a change in how we see mathematics itself. What we do now know is school mathematics is quite different from workplace mathematics. Because mathematics is "shaped" by the workplace context, rather than procedural, this leaves them unprepared for tasks in which mathematics is embedded and functional.

Class, in some guise or another, is always a latent variable whose invisibility obscures possibilities for action.

The role of class remains not merely an epistemic or empirical question, but a political and an ideological one. However, if failure in mathematics is as structured and systematic as the OECD report seems to suggest, why is this not clearer in mathematics education research? That is indeed an ideological question. Pais and Valero (2012, p. 18) argue "although many researchers acknowledge the social and political aspects involved in reforming mathematics education, they end up investigating problems as if they could be solved through better classroom practices". If we are to change things, we have to more away from claiming that such considerations are "beyond the scope of this book". We need to engage more with the consequences of the economy which structures our existence, our exchanges and our relationships. This might mean shifting away from a denial of grand narratives, and looking instead toward those structural explanations of the social world which have proved successful for almost two hundred years.

# Chapter 3
# Summary and Looking Ahead

After some four decades, the *Social and Political Dimensions of Mathematics Education* has become mainstreamed by its inclusion as a new Topic Study Group in ICME-13 for the first time. This topical survey demonstrates the diversity of scholarship and practice that has grown through five key areas that have been explored.

**Summary of findings**:

- Equitable access and participation in mathematics education, is achievable to a high a degree in some countries, such as Cuba and Finland. Ideology and theoretical perspectives shape to a great extent, the policies on equitable access and participation in mathematics education.
- It is evident that mathematics is increasing perceived as a negotiable field of social practices arising from specific needs and serving certain interests, which open or close possibilities for understanding and shaping our world.
- Extensive research on the lived experience in mathematics education has shed light on how and why students do and don't identify with mathematics. However, this research seems to re-entrench stereotypes about identities that excel at mathematics and also often falls into the trap of assuming a binary between structure and agency.
- What questions are researched, by whom, in what settings and for what purposes shapes and frames particular discourses, as they emerge, get taken up and become dominant or disappear. In this context the relations between activism, the material conditions of inequality and mathematics education has remained under-developed and under-represented.
- The nature of a society's economic structure influences not only public interactions, but also very localised social relations, including those in the classroom. The result of this is a marginalisation of learners from disadvantaged communities and specifically children in poor and working class households. Such learners suffer curriculum exclusion and an experience which places the

© The Author(s) 2016
M. Jurdak et al., *Social and Political Dimensions of Mathematics Education*,
ICME-13 Topical Surveys, DOI 10.1007/978-3-319-29655-5_3

responsibility for failure back upon the shoulders of the disadvantaged, rather than the affluent whose privilege everyone else pays for.

**Looking ahead**:

- Questions need to be asked about moving from definitions of quality of mathematics education in technical terms, independent of social context, to definitions of quality in terms of social practice that are embedded in socially constructed epistemological principles.
- Apart from gaining further insights on how mathematics educations contributes to reproducing social inequality, more research is needed on the political bias of central—and too often taken-for-granted—concepts and convictions of mathematics and mathematics education.
- Multiple, in depth case-studies are required that examine the policies, economic and material conditions, and the type of activism that are favorable to move toward more equitable access and participation in quality mathematics education.
- Most identity research draws on discourse studies of various kinds (language based). There is a great need for innovative different kinds of research methods (other than interview and survey) and different kinds of data (other than spoken or written responses) to really tap into the lived experience of mathematics students and teachers.
- Analysis of the influence of the economic superstructure upon mathematics achievement identified the extent to which income inequality affects fundamental principles of equity, social justice, and in turn achievement in mathematics. Therefore, a key strategy for those working in mathematics education concerned about levels of achievement has to be to work for a reduction in income inequality.

The crucial importance of this last area and its relevance in the current global context of rising inequality, unemployment (especially the youth) and increasing poverty may well require an acknowledgement through an explicit expansion of this Topic Study Group to a focus on the Social, Political *and Economic* Dimensions of Mathematics Education into the future.

# References

Andrade Molina, M., & Valero, P. (in press). The sightless eyes of reason: Scientific objectivism and school geometry. In N. Vondrová (Ed.), *Proceedings of the Ninth Congress of the European Society for Research in Mathematics Education*.

Askew, M., Hodgen, J., Hossain, S., & Bretcher, N. (2010). *Values and variables: Mathematics education in high-performing countries*. London: Nuffield Foundation.

Atweh, B. (2011). Quality and equity in mathematics education as ethical issues. In B. Atweh, M. Graven, W. Secada, & P. Valero (Eds.), *Mapping equity and quality in mathematics education* (pp. 63–75). Netherlands: Springer.

Balfanz, R., Mac Iver, D. J., & Byrnes, V. (2006). The implementatiom and impact of evidence-based mathematics reforms in high-poverty middle schools: A multi-site, multi-year study. *Journal for Research in Mathematics Education, 37*, 33–64.

Baxter, J. A., Woodward, J., Voorhies, J., & Wong, J. (2002). We talk about it, but do they get it? *Learning Disabilities Research & Practice, 17*, 173–185.

Bembechat, J. (1998). *Against the odds: How "at risk" students exceed expectations*. San Francisco: Jossey Bass.

Benwell, B., & Stokoe, E. (2006). *Discourse and identity*. Edinburgh University.

Bernstein, B. (1971). *Theoretical studies towards a sociology of language*. London: Routledge.

Black, L., Mendick, H., & Solomon, Y. (Eds.). (2011). *Mathematical relationships in education: Identities and participation*. Abingdon: Routledge.

Boaler, J., & Greeno, J. G. (2000). Identity, agency and knowing in mathematics worlds. In J. Boaler (Ed.), *Multiple perspectives on mathematical teaching and learning*. London: Greenwood.

Borgioli, G. M. (2008). A critical examination of learning disabilities in mathematics: Applying the lens of ableism. *Journal of Thought, 43*(12), 131–147.

Bourdieu, P. (1986). The forms of capital. In J. G. Richardson (Ed.), *Handbook of theory and research for the sociology of education* (pp. 241–260). New York: Greenwood.

Brown, T. (2010). Truth and the renewal of knowledge: The case of mathematics education. *Educational Studies in Mathematics, 75*(3), 329–343.

Brown, T., & McNamara, O. (2011). *Becoming a mathematics teacher: Identity and identification*. Dordrecht: Springer.

Brown, T., Hodson, E., & Smith, K. (2013). TIMSS mathematics has changed real mathematics forever. *For the Learning of Mathematics, 33*(2), 38–43.

Buchmann, C., & Dalton, B. (2002). Interpersonal influences and educational aspirations in 12 countries: The importance of institutional context. *Sociology of Education, 75*(2), 99–122.

Buchmann, C., & Park, H. (2009). Stratification and the formation of expectations in highly differentiated educational systems. *Research in Social Stratification and Mobility, 27*(4), 245–267.

Carnoy, M., Gove, A., & Marshall, J. (2007). *Cuba's academic advantage: Why students in Cuba do better in school*. California: Stanford University.

M. Jurdak et al., *Social and Political Dimensions of Mathematics Education*,
ICME-13 Topical Surveys, DOI 10.1007/978-3-319-29655-5

Chronaki, A. (2011). Disrupting 'development' as the quality/equity discourse: Cyborgs and subalterns in school technoscience. In B. Atweh, M. Graven, W. Secada, & P. Valero (Eds.), *Mapping equity and quality in mathematics education* (pp. 3–19). Netherlands: Springer.

Cooper, B., & Dunne, M. (2000). *Assessing children's mathematical knowledge: Social class, sex, and problem-solving.* Buckingham: Open University.

Davis, P. J., & Hersh, R. (1983). *The mathematical experience.* Harmondsworth: Penguin.

de Freitas, E. (2004). Plotting intersections along the political axis: The interior voice of dissenting mathematics teachers. *Educational Studies in Mathematics, 55,* 259–274.

de Freitas, E. (2008). Troubling teacher identity: Preparing mathematics teachers to teach for diversity. *Teaching Education, 19*(1), 43–55.

de Freitas, E. (2008a). Mathematics and its other: (Dis)locating the feminine. *Gender and Education, 20*(3), 281–290.

de Freitas, E. (2013). The mathematical event: Mapping the axiomatic and the problematic in school mathematics. *Studies in Philosophy and Education, 32*(6), 581–599.

de Freitas, E., & Sinclair, N. (2014). *Mathematics and the body: Material entanglements in the classroom.* Cambridge University.

Desrosières, A. (1993). *La politique des grands nombres: Histoire de la raison statistique. [The politics of large numbers: History of statistical thinking.]* Paris: La Découverte.

Dowling, P. (1998). *The sociology of mathematics education: Mathematical myths/pedagogic texts.* London: Falmer.

Drake, C., Spillane, J. P., & Hufferd-Ackles, K. (2001). Storied identities: Teacher learning and subject-matter context. *Journal of Curriculum Studies, 33*(1), 1–23.

Engeström, Y. (1987). *Learning by expanding: an activity-theoretical approach to developmental research.* Helsinki: Orienta-Konsultit.

Engeström, Y. (2001). Expansive learning at work: Toward an activity theoretical reconceptualization. *Journal of Education & Work, 14*(1), 133–156.

Esmonde, I. (2012). Mathematics learning in groups: Analyzing equity within an activity structure. In B. Herbel-Eisenmann, J. Choppin, D. Wagner, & D. Pimm (Eds.), *Equity in discourse for mathematics education.* Dordrecht: Springer.

Evans, J., Tsatsaroni, A., & Czarnecka, B. (2014). Mathematical images in advertising. *Educational Studies in Mathematics, 85*(1), 3–27.

Foucault, M. (1984). Truth and power. In P. Rabinow (Ed.), *The Foucault reader* (pp. 51–75). New York: Pantheon.

Fuchs, L. S., Fuchs, D., Hamlett, C. L., & Appleton, A. C. (2002). Explicitly teaching for transfer: Effects on the mathematical problem-solving performance of students with mathematics disabilities. *Learning Disabilities Research & Practice, 17,* 90–106.

Gasperini, L. (2000). *The Cuban educational system: Lessons and dilemmas.* Washington, D. C.: The World Bank.

Gates, P., & Cotton, T. (1998). *Proceedings of the First International Mathematics Education and Society Conference.* Nottingham University.

Gates, P., & Guo, X. (2013). How British-Chinese parents support their children: A view from the regions. *Educational Review, 66*(2), 168–191.

Gates, P., & Vistro-Yu, C. (2003). Is mathematics for all? In A. Bishop, M. Clements, C. Keitel, J. Kilpatrick, & F. Leung (Eds.), *Second international handbook of mathematics education* (pp. 31–73). Dordrecht: Kluwer.

Gifford, S., & Rockliffe, F. (2008). In search of dyscalculia. *Proceedings of the British Society for Research into Learning Mathematics, 28*(1), 21–27.

Graven, M. H. (2014). Poverty, inequality and mathematics performance: The case of South Africa's post-apartheid context. *ZDM, 46,* 1039–1049.

Greer, B. (2012). The USA mathematics advisory panel: A case study. In B. Greer & O. Skovsmose (Eds.), *Opening the cage: Critique and politics of mathematics education* (pp. 107–125). Rotterdam: Sense.

Greer, B., & Skovsmose, O. (2012). Seeing the cage? The emergence of Critical Mathematics Education. In B. Greer & O. Skovsmose (Eds.), *Opening the cage: Critique and politics of mathematics education* (pp. 1–20). Rotterdam: Sense.

Griffiths, T. G. (2009). Fifty years of socialist education in revolutionary Cuba: A world-systems perspective. *Journal of Iberian and Latin American Research, 15*(2), 45–64.

Gutiérrez, R. (1999). Advancing urban Latina/o youth in mathematics: Lessons from an effective high school mathematics department. *Urban Review, 31*(3), 263–281.

Gutiérrez, R. (2008). A 'gap-gazing' fetish in mathematics education? Problematizing research on the achievement gap. *Journal for Research in Mathematics Education, 39*(4), 357–364.

Gutiérrez, R. (2013). The sociopolitical turn in mathematics education. *Journal of Research in Mathematics Education, 4*(1), 37–68.

Gutiérrez, R., & Dixon-Román, E. (2011). Beyond gap gazing: How can thinking about education comprehensively help us (re)envision mathematics education? In B. Atweh, M. Graven, W. Secada, & P. Valero (Eds.), *Mapping equity and quality in mathematics education* (pp. 21–34). Netherlands: Springer.

Gutstein, E. (2012). Mathematics as a weapon in the struggle. In B. Greer & O. Skovsmose (Eds.), *Opening the cage: Critique and politics of mathematics education* (pp. 23–48). Rotterdam: Sense.

Gutstein, E. (2003). Teaching and learning mathematics for social justice in an urban, Latino school. *Journal for Research in Mathematics Education, 34*(1), 37–73.

Hall, S. (1996). Who needs identity? In S. Hall & P. Du Gay (Eds.), *Questions of cultural identity* (pp. 1–17). London: Sage.

Healy, L., & Powell, A. (2013). Understanding and overcoming 'disadvantage' in learning mathematics. In M. A. Clements, A. Bishop, C. Keitel-Kreidt, J. Kilpatrick, & F. K.-S. Leung (Eds.), *Third international handbook of mathematics education* (pp. 69–100). Berlin: Springer.

Healy, L., Hassan, S., & Fernandes, A. A. (2011). The role of gestures in the mathematical practices of those who do not see with their eyes. *Educational Studies in Mathematics, 77*, 157–174.

Hehir, T. (2005). *New directions in special education: Eliminating ableism in policy and practice.* Cambridge University Press.

Hoyles, C., Noss, R., Kent, P., & Bakker, A. (2010). *Improving mathematics at work: The need for techo-mathematical literacies.* London: Routledge.

Hursh, D. (2007). Assessing No Child Left Behind and the rise of neoliberal education policies. *American Educational Research Journal, 44*(3), 493–518.

Jorgensen, R., Gates, P., & Roper, V. (2014). Structural exclusion through school mathematics: Using Bourdieu to understand mathematics a social practice. *Educational Studies in Mathematics, 87*, 221–239.

Julie, C., Angelis, D., & Davis, Z. (Eds.). (1993). *Political dimensions of mathematics education: Curriculum reconstruction for society in transition.* Cape Town: Maskew Miller Longman.

Jurdak, M. (2009). *Equity in quality in mathematics education.* New York: Springer.

Jurdak, M. (2014). Socio-economic and cultural mediators of mathematics achievement and between-school equity in mathematics education at the global level. *ZDM Mathematics Education, 46*(7), 1025–1037.

Kanes, C., Morgan, C., & Tsatsaroni, A. (2014). The PISA mathematics regime: Knowledge structures and practices of the self. *Educational Studies in Mathematics, 87*(2), 145–165.

Keitel, C. (Ed.). (1989). *Mathematics, education and society.* Paris: UNESCO.

Kitchen, R. (2003). Getting real about mathematics education reform in high-poverty communities. *For the Learning of Mathematics, 23*, 16–22.

Kjaergard, T., Kvamme, A., & Linden, N. (1995). *The Third International Conference on the Political Dimensions of Mathematics Education: Numeracy, race, gender, and class.* Landås: Caspar.

Knijnik, G. (1999). Ethnomathematics and the Brazilian landless people education. *ZDM Mathematics Education, 99*(3), 96–99.

Kohn, M. (1983). On the transmission of values in the family: A preliminary foundation. *Research in the Sociology of Education and Socialisation, 4*(1), 1–12.

Kollosche, D. (2014). Mathematics and power: An alliance in the foundations of mathematics and its teaching. *ZDM Mathematics Education, 46*(7), 1061–1072.

Lee, J. (2012). Educational equity and adequacy for disadvantaged minority students: School and teacher resource gaps toward national mathematics proficiency standards. *The Journal of Educational Research, 105,* 64–75.

Leont'ev, A. N. (1981). The problem of activity in psychology. In J. V. Wertsch (Ed.), *The concept of activity in Soviet psychology* (pp. 37–71). New York: Taylor & Francis.

Lerman, S. (2000). The social turn in mathematics education research. In J. Boaler (Ed.), *Multiple perspectives on mathematics teaching and learning* (pp. 19–44). Westport, CT: Ablex.

Lerman, S. (2014). Mapping the effects of policy on mathematics teacher education. *Educational Studies in Mathematics, 87*(2), 187–201.

Llewellyn, A. (2012). Unpacking understanding: The (re)search for the Holy Grail of mathematics education. *Educational Studies in Mathematics, 81*(3), 385–399.

Lubienski, S. (2000a). A clash of cultures? Students' experiences in a discussion-intensive seventh grade mathematics classroom. *Elementary School Journal, 100,* 377–403.

Lubienski, S. (2000b). Problem solving as a means towards mathematics for all: An exploratory look through a class lens. *Journal for Research in Mathematics Education, 31*(4), 454–482.

Lubienski, S. (2002). Good intentions were not enough: Lower SES students' struggles to learn mathematics through problem solving. In J. Sowder & B. Schapelle (Eds.), *Lessons learned from research* (pp. 171–178). Reston, VA: NCTM.

Lubienski, S., & Gutiérrez, R. (2008). Bridging the gaps in perspectives on equity in mathematics education. *Journal for Research in Mathematics Education, 39*(4), 365–371.

Lundin, S. (2012). Hating school, loving mathematics: On the ideological function of critique and reform in mathematics education. *Educational Studies in Mathematics, 80*(1–2), 73–85.

Martin, D. B. (2009). Researching race in mathematics education. *Teacher College Record, 111,* 295–338.

Marx, K. (1972). The German ideology. In R. C. Tucker (Ed.), *The Marx-Engels reader* (pp. 146–200). New York: Norton.

McKinney, S., & Frazier, W. (2008). Embracing the Principles and Standards for School Mathematics: An inquiry into the pedagogical and instructional practices of mathematics teachers in high-poverty middle schools. *Clearing House, 81,* 201–210.

Mellin-Olsen, S. (1987). *The politics of mathematics education.* Dordrecht: Reidel.

Mendick, H. (2006). *Masculinities in mathematics.* London: Open University Press.

Monseur, C., & Lafontaine, D. (2012). Structure des systèmes éducatifs et équité: Un éclairage international [Structures of education systems and equity: An international perspective.]. In M. Crahay (Ed.), *Pour une école juste et efficace [Towards a more effective and equitable school.].* Brussels: De Boeck.

Morgan, H. (2014). The education system in Finland: A success story other countries can emulate. *Childhood Education, 90*(6), 453–457.

Noss, R., Brown, A., Dowling, P., Drake, P., Harris, M., Hoyles, C., & Mellin-Olsen, S. (1990). *Political dimensions of mathematics education: Action and critique.* University of London.

Nunes, T. (2004). *Teaching mathematics to deaf children.* London: Whurr.

OECD. (2013a). *PISA 2012 results: Excellence through equity: Giving every student the chance to succeed.* Paris: OECD.

OECD. (2013b). *PISA 2012 results: What makes schools successful? Resources, policies and practices.* Paris: OECD.

OECD. (2014). *PISA 2012 results: What students know and can do: Student performance in mathematics, reading and science.* Paris: OECD.

Pais, A. (2013). An ideology critique of the use value of mathematics. *Educational Studies in Mathematics, 84,* 15–34.

Pais, A. (2014). Economy: The absent centre of mathematics education. *ZDM, 46,* 1085–1093.

Pais, A., & Valero, P. (2011). Beyond disavowing the politics of equity and quality in mathematics education. In B. Atweh, M. Graven, W. Secada, & P. Valero (Eds.), *Mapping equity and quality in mathematics education* (pp. 35–48). Netherlands: Springer.

Pais, A., & Valero, P. (2012). Researching research: Mathematics education in the political. *Educational Studies in Mathematics, 80*, 9–24.

Payne, K. J., & Biddle, B. J. (1999). Poor school funding, child poverty, and mathematics achievement. *Educational Researcher, 28*, 4–13.

Pickett, K., & Wilkinson, R. (2015). Income inequality and health: A causal review. *Social Science and Medicine, 128*, 316–326.

Popkewitz, T. (2002). Whose heaven and whose redemption? In P. Valero & O. Skovsmose (Eds.), *Third International Mathematics Education and Society Conference* (pp. 1–26). Helsingør: Centre for Research in Learning Mathematics.

Porter, T. M. (1996). *Trust in numbers: The pursuit of objectivity in science and public life.* Princeton University Press.

Radford, L. (2003). On the epistemological limits of language: Mathematics knowledge and social practice during the renaissance. *Educational Studies in Mathematics, 52*, 123–150.

Radford, L. (2012). Education and the illusions of emancipation. *Educational Studies in Mathematics, 80*, 101–118.

Rodríguez, A. J., & Kitchen, R. S. (Eds.). (2005). *Preparing mathematics and science teachers for diverse classrooms: Promising strategies for transformative pedagogy.* Mahwah, NJ: Lawrence Erlbaum.

Roth, W.-M. (2014). Rules of bending, bending the rules: The geometry of electrical conduit bending in college and workplace. *Eduational Studies Mathematics, 86*, 117–192.

Sarjala, J. (2013). Equality and cooperation: Finland's path to excellence. *American Educator, 37* (1), 32–36.

Skovsmose, O. (1994). *Towards a philosophy of critical mathematics education.* Dordrecht: Kluwer.

Skovsmose, O. (2011). *An invitation to critical mathematics education.* Rotterdam: Sense.

Solomon, Y. (2009). *Mathematical literacies: Developing identities of inclusion.* New York: Routledge.

Stinson, D. W. (2004). Mathematics as 'gate-keeper'. *The Mathematics Educator, 14*(1), 8–18.

Stinson, D. W. (2013). Negotiating the 'white male math myth': African American male students and success in school mathematics. *Journal for Research in Mathematics Education, 44*(1), 69–99.

Stotesbury, N., & Dorling, D. (2015). Understanding income inequality and its mplications: Why better statistics are needed. Resource document. Statistics Views. Retrieved October 21, 2015, from http://www.statisticsviews.com/details/feature/8493411/Understanding-Income-Inequality-and-its-Implications-Why-Better-Statistics-Are-N.html.

Straehler-Pohl, H., Fernández, S., Gellert, U., & Figueiras, L. (2014). School mathematics registers in a context of low academic expectations. *Educational Studies in Mathematics, 85*(2), 175–199.

Tracy, K. (2002). *Everyday talk: Building and reflecting identities.* New York: Guilford.

Tsatsaroni, A., & Evans, J. (2014). Adult numeracy and the totally pedagogised society: PIAAC and other international surveys in the context of global educational policy on lifelong learning. *Educational Studies in Mathematics, 87*(2), 167–186.

Turner, S. E. (2000). A comment on "Poor School Funding, Child Poverty, and Mathematics Achievement". *Educational Researcher, 29*, 15–18.

Ullmann, P. (2008). *Mathematik, Moderne, Ideologie: Eine kritische Studie zur Legitimität und Praxis der modernen Mathematik* [Mathematics, modernity, ideology: A critical study on the legitimacy and praxis of modern mathematics.]. Konstanz: UVK.

Valero, P. (2004). Socio-political perspectives on mathematics education. In P. Valero & R. Zevenbergen (Eds.), *Researching the socio-political dimensions of mathematics education. Issues of power in theory and methodology* (pp. 5–23). Boston: Kluwer.

Valero, P., & Zevenbergen, R. (Eds.). (2004). *Researching the socio-political dimensions of mathematics education. Issues of power in theory and methodology*. Boston: Kluwer.

Valero, P., García, G., Camelo, F., Mancera, G., & Romero, J. (2012). Mathematics education and the dignity of being. *Pythagoras, 33*(2).

Valero, P., Graven, M., Jurdak, M., Martin, D., Meaney T., & Penteado, M. (2015). Socioeconomic influence on mathematical achievement: What is visible and what is neglected. In S. J. Cho (Ed.), *The proceedings of the 12th International Congress on Mathematical Education* (pp. 285–301).

Vijay, R., Zuze, T. L., Visser, M., Winaar, L., Juan, A., Prinsloo, C. H., et al. (2015). *Beyond benchmarks: What twenty years of TIMSS data tell us about South African education*. Cape Town: HSRC.

Vithal, R. (2003). *In search of a pedagogy of conflict and dialogue for mathematics education*. Dordrecht: Kluwer.

Vithal, R. (2012). Mathematics education, democracy and development: Exploring connections. *Pythagoras, 33*(2), 20–33.

Walkerdine, V. (1998). *Counting girls out: Girls and mathematics*. London: Falmer.

Walshaw, M. (2001). A Foucauldian gaze on gender research: What do you do when confronted with the tunnel at the end of the light? *Journal for Research in Mathematics Education, 32*(5), 471–492.

Walshaw, M. (Ed.). (2004). *Mathematics education within the postmodern*. Greenwich, CO: Information Age.

Walshaw, M. (2013). Explorations into pedagogy within mathematics classrooms: Insights from contemporary inquiries. *Curriculum Inquiry, 43*(1), 71–94.

Wetherell, M. (1998). Positioning and interpretative repertoires: Conversation analysis and post-structuralism in dialogue. *Discourse & Society, 9*(3), 387–412.

Widdicombe, S. (1998). Identity as an analysts' and a participants' resource. In C. Antaki & S. Widdicombe (Eds.), *Identities in talk* (pp. 191–206). London: SAGE.

Wilkinson, R. (2005). *The impact of inequality*. London: Routledge.

Wilkinson, R., & Pickett, K. (2010). *The spirit level*. London: Allen Lane.

Woodward, J., & Montague, M. (2002). Meeting the challenge of mathematics reform for students with learning difficulties. *Journal of Special Education, 36*(2), 89–101.

Wortham, S. E. F. (2006). *Learning identity: The joint emergence of social identification and academic learning*. Cambridge University Press.

Zevenbergen, R. (1996). Constructivism as a liberal bourgeois discourse. *Educational Studies in Mathematics, 31*, 95–113.

Zevenbergen, R. (2000). "Cracking the code" of mathematics classrooms: School success as a function of linguistic, social and cultural background. In J. Boaler (Ed.), *Multiple perspectives on mathematics teaching and learning* (pp. 201–223). Westport: Ablex.

Zevenbergen, R. (2005). The construction of a mathematical habitus: Implications of ability grouping in the middle years. *Journal of Curriculum Studies, 37*(5), 607–619.

## Further Reading

The proceedings of the international Mathematics Education and Society (MES) conferences provide a useful indication of how the social and political dimensions of mathematics education have grown and the diversity of areas and issues being engaged.

(The MES conference proceedings can be accessed at: www.mescommunity.info)

Mukhopadhyay, S., & Greer, B. (Eds.) (2015). *Proceedings of the Eighth International Mathematics Education and Society Conference*. Portland: Ooligan.

Berger, M., Brodie, K., Frith, V., & le Roux, K. (Eds.) (2013). *Proceedings of the Seventh International Mathematics Education and Society Conference*. Cape Town: Hansa.

Gellert, U., Jablonka, E., & Morgan, C. (Eds.). (2010). *Proceedings of the Sixth International Mathematics Education and Society Conference.* Freie Universität Berlin.

Matos, J. F., Valero, P., & Yasukawa, K. (Eds.). (2008). *Proceedings of the Fifth International Mathematics Education and Society Conference.* Universidade de Lisboa.

Goos, M., Kanes, C., & Brown, R. (Eds.) (2004). *Proceedings of the 4th International Mathematics Education and Society Conference.* Griffith University.

Valero, P., & Skovsmose, O. (Eds.). (2005). *Proceedings of the Third International Mathematics Education and Society Conference.* Danmarks Pædagogiske Universitet.

Matos, J. F., & Santos, M. (Eds.). (2000). *Second International Mathematics Education and Society Conference. Proceedings.* Universidade de Lisboa.

Gates, P., & Cotton, T. (Eds.). (1998). *Proceedings of the First International Mathematics Education and Society Conference.* University of Nottingham.